Mrs. Lindh

Wonderful Water

Acknowledgments
Executive Editor: Diane Sharpe
Supervising Editor: Stephanie Muller
Design Manager: Sharon Golden
Page Design: Simon Balley Design Associates
Photography: Norman MacBeath: page 25; Oxford Scientific Films:
cover (top, left), pages 9, 13, 17, 21, 27; Alex Ramsay: page 7;
Tony Stone: cover (bottom).

ISBN 0-8114-3771-X

Wonderful
Water

Rosie Hankin

Illustrated by
Stuart Trotter

STECK-VAUGHN
C O M P A N Y
ELEMENTARY • SECONDARY • ADULT • LIBRARY

All plants need water so they can grow.

Grass does not grow if there is too little
rain. Then there is not enough for
animals, such as cows and sheep, to eat.
All animals need water.

7

Rainwater that does not soak into the ground runs off into streams. When there has been a heavy rainfall, streams flow very fast.

Then the small streams join bigger streams that become rivers.

The rain stopped.

Look at the boats on that huge lake!

That's not a lake. It's a reservoir.

10

A reservoir is a place where water
is collected. Then the water is pumped
out to be used for other things.

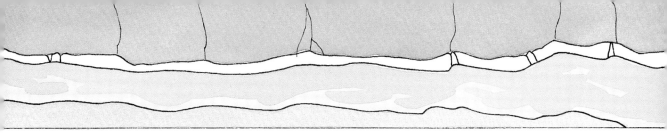

A dam controls the amount of water
that flows out of the reservoir. Huge
underwater gates are opened and
closed, depending on how much water
is needed.

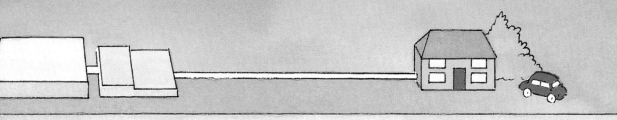

When water is allowed to flow out of the dam, it is pumped through pipes to many other places. It is pumped to power stations, factories, and even houses.

Inside a hydroelectric power station, there are huge machines called turbines. Water that flows from the reservoir spins the turbines to make electricity.

Some water from the reservoir is cleaned at a filtration plant like this one. Mud is filtered out, and germs are killed. Then the clean water is pumped to houses and factories.

Deer, foxes, and birds come to the river to drink. They need water to survive.

When we are hot, our skin sweats. We need to drink liquids to replace the water we lose as sweat.

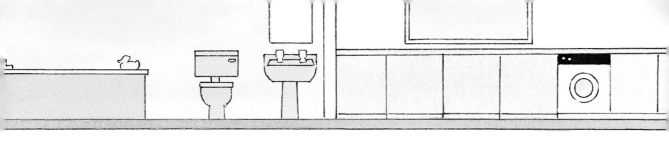

We use water in our homes for washing, cooking, and cleaning.

25

Water is cleaned at a sewage plant. Then the clean water is pumped back to houses and factories. Sometimes it is pumped into rivers or lakes.

I think it's wonderful!

Water is very important. All living things need it to survive. But people use water in many other ways, too. We use it for cooking, washing, and even making electricity.

Look at the pictures. Do you remember the different ways that water is used? The answers are on the last page, but don't look until you have tried naming everything.

1.

2.

3.

4.

5.

6.

7.

Answers: 1. For animals to drink 2. For making electricity
3. For people to drink 4. For brushing teeth 5. For watering plants
6. For cooking 7. For boats